An ancient olive tree from the Alantejo, Portugal.
(Photo by Nicolas Sapieha)

for Gregory, Henry, Sophia and Felix

Author's Acknowledgments

I am grateful to Maria Teresa Train for major research, and to Francesca Manisco, for editorial assistance. Also to Lisa Train Pinnington and Sara Perkins, for reviewing the text.

J.T.

Design by: Natasha Tibbott, Our Designs, Inc.

Photolithos and Printing: Sfera, Milano

ISBN: 1-85149-473-1
L.o.C.: 2004105469

PREVIOUS PAGE: *Olive grove in Corfu. Photographed 1880.*
(PHOTO FROM FRATELLI ALINARI SPA, FLORENCE)

For Andy
from
(uncle) John

John Train

The Olive
Tree of Civilization

Olive leaves from a victory crown

PHOTOGRAPHY BY
Mark E. Smith

DISTRIBUTED BY
ANTIQUE COLLECTORS' CLUB

EASTHAMPTON, MA

WOODBRIDGE, U.K.

The olive tree is surely
the richest gift of heaven.

THOMAS JEFFERSON

Oil is a symbol of God, for it can heal like His mercy and sear like His anger, and is itself colourless and cold, yet has all colour and heat within it, as God, who has no human attributes, is the source of all human attributes. Oil has had many loving names conferred on it: the holy oil, the oil of gladness, the oil of sanctification, a royal robe, the delight of the heart, an eternal joy, the oil of salvation.

REBECCA WEST

Engraving of a bull with an olive wreath, inspired by Ovid's Metamorphoses.

Contents

Origins

The olive is intimately linked with the history of mankind and the growth of western civilization. It has been called the tree of civilization, because unlike raw crops planted in spring, which yield production each season, it must be tended for years before bearing fruit. In a politically disordered environment the farmer may well shrink from so long-term a commitment. Thus, Shakespeare, in *Antony and Cleopatra*, "The time of universal peace is near. Prove this a prosp'rous day, the three-nook'd world shall bear the olive freely."

The oldest known locus of the wild olive tree goes back over 40 millennia, in the Sahara, which at that time was fertile. (Even today, there lies beneath its sands a vast aquifer.) The earliest indication we have of systematic cultivation of the olive dates back 8,000 years, to ancient Egypt. Inevitably, this useful technique radiated around the region, presumably carried by those great traders of antiquity, the Phoenicians. Groves in Crete date back about 5,000 years, and production in Asia Minor is of comparable antiquity.

Chiseled gold cup from Vafio, Greece. XV century B.C.

For the ancient Egyptians, Isis, goddess of the moon and light, and wife of Osiris, taught them the olive's cultivation and its many uses.

For the Mediterranean peoples, the olive serves an extraordinary variety of purposes: food (both fruit and oil), medicine, cosmetics, illumination, fodder for the livestock, and a superior wood for furniture and construction, and firewood for the house. Except for the palm in the Pacific, which also provides building material, wine and even a kind of paper, nothing equals the olive's many rich uses.

Trade in olives was regulated as far back as 2,500 B.C., under the rule of Hammurabi. The statesman and poet Solon (640–c. 558 B.C.) promulgated regulations governing its cultivation in Attica.

The Greeks held that the olive tree was sacred, and sometimes punished its improper cutting by death, banishment, or confiscation of lands. At one time, harvesting the olive was limited to virgin girls and youths sworn to chastity.

Cameo in onyx and sardonyx, depicting Poseidon and Athena competing for dominion over Attica. Athena wins by creating the olive tree. First century B.C. *Archeological Museum, Naples.*
(PHOTO BY ALFREDO AND PIO FOGLIA, NAPLES)

We find Egyptian bas-reliefs depicting olive culture, as well as Minoan frescos, Greek jars, Roman silver vases, and Carthaginian mosaics. They sometimes depict harvesting the tree by tapping the branches with sticks—a technique one still sees today.

In Greek legend, Athena (Minerva) and Poseidon (Neptune) both wanted to found the city that would be Athens. The quarrel, becoming intense, was referred to Zeus (Jupiter) king of the gods, for decision. Zeus ruled that whichever of the two claimants granted the most useful gift to the people of the future city could become its patron. Poseidon struck the earth with his trident, and a warhorse sprang up. Athena, instead, produced an olive tree, source of a hundred benefits. Zeus ruled in favor of Athena, whence the name of the city.

Red and black varnished ceramic vase, Apulia, Italy. IV century B.C. Museo dell'Olio Fondazione Lungarotti, Torgiano.

Thus, Sophocles, in *Oedipus at Colonus*, "A glorious tree flourishes in our Dorian land: Our sweet, silvered wet nurse, the olive. Self-born and immortal, unafraid of foes, her ageless strength defies knaves young and old, for Zeus and Athena guard her sleepless eyes."

In the era of the Peloponnesian Wars, military technique favored the defense. A town could successfully withstand a siege behind its strong fortifications. The invaders might lay waste to the surrounding fields, and put the torch to the olive groves.[1] But surprise! That hardy tree, scorched, would still put out new buds the next season. So to destroy a grove an invader had to dig up the trees by the roots. Since a warrior hates to put aside his sword, and with it the hope of glorious deeds, to spend the campaigning season with a pick and shovel, the olive groves usually survived.

The Greeks established prosperous colonies in Sicily between 800 and 500 B.C., bringing the cultivation of the olive with them, which thence moved to Italy proper. (Roman *olea* derives from Greek *elea*.) From there it spread to France, Spain, and North Africa, where the Romans planted extensively. In later centuries the Arab invaders of North Africa, who later conquered much of Spain, brought their own words for the olive and its oil—*aceituna* and *aceite*—into the Spanish language.

---

[1] Judges 15:57x describes how Samson, warring with the Philistines, "released hordes of foxes with torches tied to their tails. The terrified foxes burnt up both the shocks (of corn) and also the standing corn, with the vineyards and olives."

Antique oil amphora by the Painter of Priano, Cerveteri, Italy. Villa Giulia, Rome.
(Photo from Sopraintendenza per i Beni Archeologici)

Soldiers, like sailors, should be kept busy in peacetime, either in martial exercises or useful work. That most prodigious of generals, Hannibal, is believed to have occupied his soldiers, when they were not in the field, with planting olive groves; as a result, Carthage became a major producer. According to the Latin author Juvenal the Carthaginian oil that flooded into Rome was rated quite unfit for human consumption, only for burning in lamps.

The Romans constructed a hundred-mile aqueduct from Jebal to Laptis Magna, in Libya. Olive oil was poured onto the water being transported at Jebal, and skimmed off again in Laptis to be exported.

A touching passage in Ovid's *Metamorphoses* describes Baucis and Philemon, poor peasants, receiving a pair of unknown visitors (who in fact are Jupiter and Mercury, disguised).

> Baucis, her skirts tucked up, was setting the table
> With trembling hands. One table leg was wobbly;
> A piece of shell fixed that. She scoured the table,
> Made level now, with a handful of green mint,
> Put on the olives, black or green, and cherries
> Preserved in dregs of wine, endive and radish,
> And cottage cheese, and eggs, turned over lightly
> In the warm ash, with shells unbroken.

The culture of the olive reached Armenia and Persia early, and spread to the Caspian region, to Afghanistan, and to China. It later flourished in Australia and South Africa.

In the sixteenth century, the olive made its way from the old world to the new, borne by Spanish missionaries, who eventually brought it to America, starting with Peru, Chile, Argentina, and Mexico. The Franciscans planted the first North American trees in about 1769, at their mission in San Diego[2], probably from Mexican seeds they had brought with them. One variety whose oil is particularly esteemed is known as the "Mission Olive;" probably all of that variety in California descend from the trees that the Franciscans planted in San Diego so long ago.

---

[2] Incidentally, the Spanish habitually named their discoveries after the saint's day on which they occurred, whence San Diego, San Francisco, and so on.

Chiseled silver cup, Pompeii. Archeological Museum, Naples.
(Photo by Alfredo and Pio Foglia)

A bronze shield with a double wreath of olive branches. Pompeii, Italy. Archeological Museum, Naples.
(Photo by Alfredo and Pio Foglia)

A Crown of Wild Olive

In rituals, the leaves and the oil of the olive have from ancient times held an honored place, signifying victory, prosperity, peace, and regeneration. An olive wreath crowned the winning athlete in the Olympic games[3], as well as a triumphant general and some of his men, a ruler, or a distinguished citizen. Cyril of Alexandria records that a herald carried olive oil to a victor in battle.

Homer describes warriors anointing themselves with olive oil after their bath. Greek and Roman athletes were massaged with olive oil before and after their exertions. The excess was then scraped off using a narrow curved instrument with an L-shaped handle called a strigil. (See page 38) The ordinary Roman was anointed and scraped after visiting the baths. A traditional late Roman definition of a pleasant life was "wine within and oil without." (In contemporary Turkey you can see heavily oiled wrestlers or *pehlivan* struggling in matches called *kirkpinar*. The winner must pin his rival's shoulders to the mat, or hurl him sideways, or—hard to do!—bear him aloft for three steps.)

Terracotta head crowned with a gold wreath encrusted with precious stones. Villa Giulia, Rome.
(Photo from Sopraintendenza per i Beni Archeologici)

Detail of a gold wreath with olive leaves, Cerveteri, Italy. Villa Giulia, Rome.
(Photo from Sopraintendenza per i Beni Archeologici)

[3] In the Pythian games, laurel, and the Nemean games, parsley.

A gold olive wreath composed of two olive branches, c. 350 B.C., Russia The Hermitage Museum, St. Petersburg.

Woman Wearing an olive crown.

Detail of the "Allegory of Good Government" depicting Peace wearing an olive crown and carrying an olive branch, by Ambrogio Lorenzetti, fresco painted 1339. Siena, Italy.
(PHOTO FROM FRATELLI ALINARI SPA, FLORENCE)

PAX

During the quadrennial panathenaic festivals, olive-crowned maidens carried offerings to the Acropolis. These nine-day festivities included competitions in music and literature, chariot and horse races, and athletic contests. The winners in the games received oil from the sacred enclosure of Athena.

In the cycle of family life, an Athenian bride was adorned with an olive garland, signifying fruitfulness, Homer mentions that she and her baby received an olive branch, which was then hung outside the house. Odysseus made Penelope's marriage-bed within the living trunk of a huge olive tree. When, upon his return from Troy, she tested his knowledge of this bed, his description if it convinced her that he was indeed her husband. A funeral in the classical period included an olive vase or *lekytos* by the body. Funerary wreaths have been found in tombs in Macedonia, South Italy, Asia Minor and the North Pontic Region. They became a popular burial offering in the fourth century B.C.

In all this the Greeks were only following their predecessors, the Egyptians, who employed olive oil in their rituals, during which participants were bathed in the oil before approaching the divinities. Olive leaves are found beside mummies, and Tutankhamen's tomb contains a silver goblet embellished with olive branches.

The Romans accorded the olive branch respect. The olive branch was a symbol of peace in ancient Rome, as elsewhere. Ambassadors seeking peace frequently carried olive branches on their hands. The Aenead describes the Trojans sailing up the Tiber to Pallanteum, where Aeneas seeks and achieves a treaty with King Evander. Standing in the prow of his vessel he offers an olive branch to Evander's son, behind whom rise the walls and buildings of the city that will become Rome. "To extend the olive branch" has thus from immemorial times been taken as a peace offering. For example, after the Battle of Bunker Hill in July 1775. Congress drew up the "Olive Branch Petition" in the hope of arranging peace with Britain. Lord North's conciliatory proposals having been refused by the colonists, the King rejected the petition.

"The Annunciation to the Shepherds," oil painting by Pietro di Sano, 1406–81. Pinacoteca, Siena.
(PHOTO FROM FRATELLI ALINARI SPA, FLORENCE)

The Dove Returned: It Found No Resting Place

A wonderful passage in Genesis describes Noah's Ark coming to rest on Mount Ararat as the flood abated. Noah releases a raven to survey the landscape. It does not reappear. After a time, Noah sends forth a dove. It returns; sent a second time, it brings back an olive leaf in its beak. Encouraging! And, released a third time, it stays away, having found a new home. Noah, taking this as a sure sign that the waters have receded, releases his family and his congeries of birds and beasts. They disembark, pairing off two by two, and the world is repopulated.

Actually, the Bible contains any number of references to the olive. Moses, in Deuteronomy, tells his people, "The Lord thy God bringeth thee into a good land, a land of brooks of water, of fountains and depths that spring out of valleys and hills; a land of wheat, and barley, and vines, and fig trees, and pomegranates, a land of olive and honey."

Later, the people are told to keep the olive for food: "Thou shalt have olive trees throughout all thy coasts, but thou shalt not anoint thyself with the oil, for thine olive shall cast his fruit."

Solomon's temple, in 1 Kings and in 2 Chronicles, contains ten-cubit cherubim of olive wood. The doors to the oracle are of olive, as are the posts of the temple doors. Psalm 128 assures whoever fears God and walks in His ways that his wife shall be as a fruitful vine, and his "children like olive plants" around his table. In all, the Bible refers to olive oil 140 times, and to the tree almost a hundred.

The Koran calls God "the light of the heaven and the earth. His light may be compared to a niche that enshrines a lamp…lit from a blessed olive tree (whose) very oil would almost shine forth, though no fire touched it." A prayer rug usually depicts an arch, or *mihrab*, representing the central niche on the mosque wall facing Mecca. In the *mihrab* is often shown a lamp, representing the burning ardor of the saints. Islam called the olive the Tree of Blessing, bringing to the world the light of Allah.

Noah, detail from the fresco "Diluvio Universale" (Noah's Flood), by Piero di Puccio, 14th century. Camposanto di Pisa, Italy.
(Photo from Fratelli Alinari SPA, Florence)

Cross in silver with a dove bearing an olive branch.

The end of Ramadan is signalized by the consumption of olive oil, and Hanukkah, in the pre-Macabees era, celebrated the end of the olive harvest. Later came the miracle of the oil lasting so many days. John Nathan, in "Jewish Cooking in America," notes that the significance of the *latkes*, pancakes, consumed in the Hanukkah celebration is that they were originally made from grain cooked in olive oil.

One remembers the Mount of Olives in the New Testament, one of the few firmly located spots in the Jerusalem area. This most ancient olive grove at the foot of the mountain is in the Garden of Gethsemane. Jesus came to this garden with His disciples to pray: "Jesus left the city and went, as he usually did, to the Mount of Olives." (Luke 22:39) Here He was arrested and then condemned to be crucified on an olive wood cross. Very little remains of the Garden of Gethsemane, except some ancient olive trees believed to be 2000 years old.

Most of the archaeological digs in the near east indicate that the inhabitants consumed oil extensively. During the Dark Ages, Christianity encouraged olive cultivation. The faithful used the oil in rituals, as a diet element for vegetarian hermits, and a convenient source of food in monasteries.

The Lord's Anointed

As Christianity succeeded the Roman religion, earlier rituals were carried over. Jesus was described as *Cristos*, consecrated by unction, or anointment. Christian baptism, confirmation, ordination of priests and bishops all involve unction. From the VII century on, the dying have received extreme unction. As under the Hebrew tradition, Christian rulers were consecrated with holy oil. Hebrew *masiah*, or messiah, very specifically means "the anointed." Incidentally, the Koran refers to Jesus many times as the messiah, although not exactly in the Christian sense. (And Mohammed, unlike many present-day Protestants, had no objection to the idea of the virgin birth, saying that if God had wanted to, he could have done it.)

The first king of Israel, Saul, 498, was anointed by applying olive oil to his forehead. Similarly, at the baptism of Clovis, the first king of the Franks, in 498, legend has it that a white dove fluttered down bearing a phial of oil (the *Sainte Ampoule*) for the anointing.

Ancient olive tree in the Garden of Gethsemane, Jerusalem.
(Photo by Bonfils, 1880 from Fratelli Alinari SPA, Florence.

A crucifix with a blessed olive branch distributed on Palm Sunday.

Detail of the Pope's miter with embroidered olive branches.

Carved olive wood dove holding an olive branch.

Fresco by Giotto depicting Christ's entry into Jerusalem. The people of the city are holding out olive branches and strewing them under the feet of the donkey He is riding. Scrovegni Chapel, Padua, Italy.
(Photo from Fratelli Alinari SPA, Florence)

Illumination

From ancient times—certainly from the fourth millennium B.C.—the oil of the olive has been used in lamps, along with resin, animal and vegetable fats, and berries. Lamps were usually an open container in which oil or tallow were burnt on a wick hanging over the edge. These lamps were used not only in religious rites, but also in homes and public places.

Relief panel depicting a lamp being filled with olive oil. Fontego dei Turchi, Venice.

Lamps from a bronze age grave on Naxos may be evidence of the earliest use of oil for illumination. One design of lamp was marble, with a concavity in the middle for the oil and three prongs to hold the wick.

As far back as the reign of Rameses III (1197–1165 B.C.), lamps burned olive oil from Heliopolis in honor of Ra, the sun god.

Ancient Cretan lamps are typically stone with a central depression and two wide spouts. Those of the Cyclades may represent birds or animals. Egyptian lamps have papyrus or linen wicks in cylindrical containers, while Phoenician lamps are usually shell-shaped.

The first Hellenistic lamps came from Southern Italy. Local craftsmen quickly developed molds to permit mass production. Lamps were made in clay, stone, bronze, silver, gold and glass. Decorative lamps soon followed in different shapes and designs.

ALE
RE
FLAM
MAM

A Coptic lamp in iron, Middle East, V century A.D. Museo dell'Olio Fondazione Lungarotti, Torgiano.

Two ancient Roman clay lamps.

Hanging lamp in Perro marble and iron, Attica, Greece, VII century. B.C.
Museo dell'Olio Fondazione Lungarotti, Torgiano.

Multicolored ceramic lamp in the shape of a duck.
Museo Etrusco, Carlo Farina, Orvieto, Italy.

Hanging iron lamp used on ships.
Museo dell'Olio Fondazione Lungarotti, Torgiano.

Silver lamp in the shape of a Roman satyr head.
Museo dell'Olio Fondazione Lungarotti, Torgiano.

Gilded lamp, Pompeii, Italy, approx. 60 A.D.
(Photo by Alfredo and Pio Foglia)

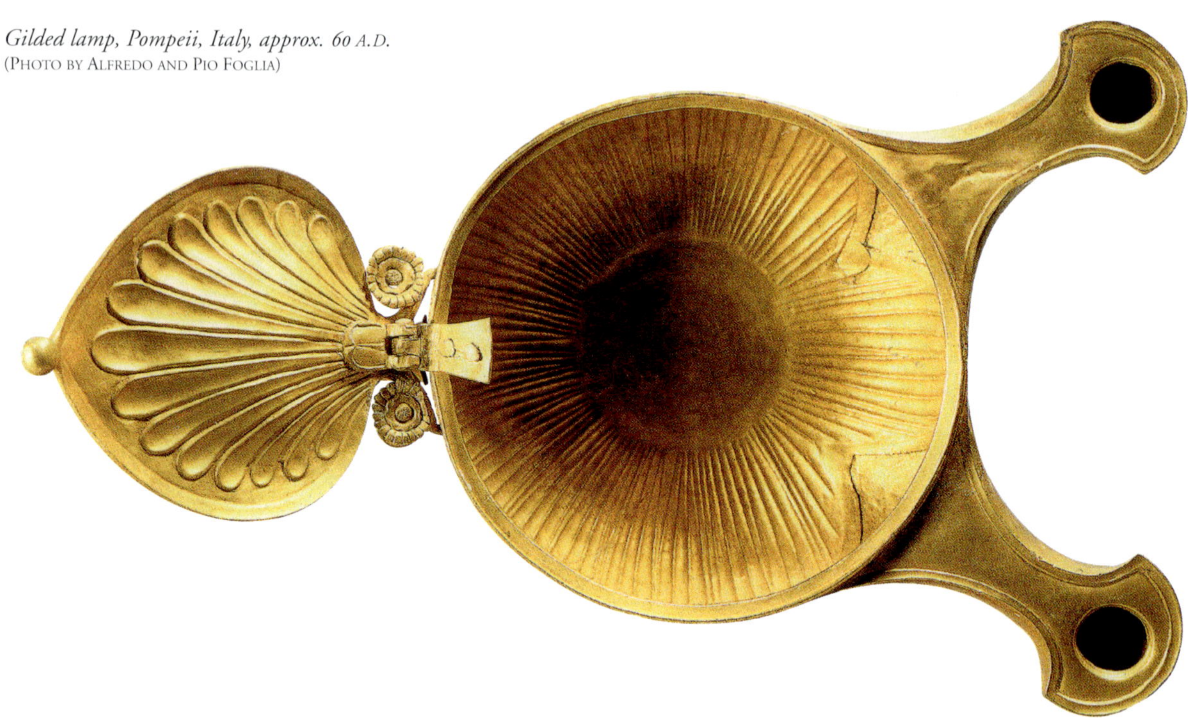

Hanukkah lamp in silver and copper, Poland. XVIII *century. Museo dell'Olio Fondazione Lungarotti, Torgiano.*

Blown glass lamp, Murano, Italy. Museo dell'Olio Fondazione Lungarotti, Torgiano.

"Ner Tanid" God's Eternal Light in silver, made for the Synagogue in Genoa, Italy, XVIII *century. Museo dell'Olio Fondazione Lungarotti, Torgiano.*

The Olive in Art

In European iconography, notably of the seventeenth and eighteenth centuries, the "Golden Age" is often represented by a woman crowned with flowers and holding an olive branch and a beehive (which, I suggest, symbolizes sweetness, industry, and light). The attributes of St. Agnes, a virgin martyr early recognized by the Church, include a lamb, the palm of martyrdom, and sometimes an olive branch or a crown of olives.

Detail from a fresco representing a religious ceremony.
Villa dei Misteri, Pompeii, Italy.
(PHOTO BY ALFREDO AND PIO FOGLIA)

In Sienese paintings the archangel Gabriel is depicted holding an olive branch, since Siena was hostile to Florence, whose emblem was the *giglio*, the lily, Gabriel's more usual offering.[4]

A dove bearing an olive branch is often used as a symbol of peace in art. Picasso's famous dove of peace unfortunately celebrates the communist version of peace, called *mir* in Russian. For the Soviets, *mir* meant stability under communist control, not everybody else's ideal.

The warm, dry, sunny landscape of Provence has always attracted painters, who have perforce depicted its characteristic tree, the olive. In 1889, Vincent Van Gogh wrote from the south, where he painted eighteen canvasses of olive trees, to his brother:

> Ah, my dear Theo, if you could see the olives at this moment... The old silver foliage and the silver-green against the blue. And the orange-hued turned earth. They are totally different from what one thinks in the north... The murmur of an olive grove has something very intimate, immensely old. It is too beautiful for me to try to conceive of it or dare to paint it.

---

[4] The angel says to Mary, "You shall conceive and bear a son, and you shall give him the name Jesus." This moment, the Annunciation, is considered to be the time of the conception of Jesus. The feast of the Annunciation, known in England as Lady Day, falls on March 25th, nine months before the nativity.

Earthenware oil jar depicting an olive branch.
Museo Etrusco Carlo Farina, Orvieto, Italy.

Jean Giono, a delightful Provencal author, wrote:

> The olive orchard is like a library, where one goes to forget life or to understand it better. In certain villages...where there is no other distraction than solitude, the men, on Sunday morning, go to the olives just as the women go to mass.

In 1936, when in Provence, Aldous Huxley wrote an essay entitled "The Olive Tree:"

> Solidified, a great fountain of life rises in the trunk, spreads in the branches, scatters in a spray of leaves and flowers and fruits. With a slow, silent ferocity the roots go burrowing down into the earth. Tender, yet irresistible, life battles with the un-living stones and has the mastery. Half hidden in the darkness, half displayed in the air of heaven, the tree stands there, magnificent, a manifest god... I like them all, but especially the olive. For what it symbolizes, first of all, peace with its leaves and joy with its golden oil.

"Cosimo I de' Medici," painting by Agnolo Bronzino, 1503–72.

Huxley claims that the ancient Hebrews loved fat, to which they ascribed religious and social significance. They wanted to anoint their kings with fat, but lacked the cows to make it possible. So they used olive oil for this purpose. He is eloquent on the olive as a painter's tree:

> Under a polished sky the olives state their aesthetic case without the qualifications of mist, of shifting lights, of atmospheric perspective, which give to the English landscape their subtle and melancholy beauty...

"The Return of Judith", tempora painting by Sandro Botticelli, 1445–1510, Uffizi Gallery, Florence. (PHOTO FROM FRATELLI ALINARI SPA, FLORENCE)

Renoir, who spent his last years in Provence, wrote:

> Look at the light on the olives, it sparkles like diamonds. It is pink, it is blue, and the sky that plays across them is enough to drive you mad.

Lawrence Durrell, in *Prospero's Cell*:

> The whole Mediterranean, the sculpture, the palms, the gold beads, the bearded heroes, the wine, the ideas, the ships, the moonlight, the winged gorgons, the bronze men, the philosophers[5]—all of it seems to rise in the sour, pungent taste of these black olives between the teeth. A taste older than meat, older than wine. A taste as old as cold water.

Rightly did Fernand Braudel, the eminent historiographer, describe the Mediterranean world as "where olives grow."

On a more vulgar level, Popeye's wife, Olive Oyl, who E.C. Segar created in 1919 for the "Thimble Theatre," is actually the oldest woman in cartoon life, predating Minnie Mouse. The Spaniards, for whom, like most Mediterranean peoples, the olive is close to a cult, did not care for that name, calling her instead "Rosario."

Mural decoration of an olive branch. Villa Pisani, Veneto, Italy.

Inlaid tabletop made of olive wood, 19th century, Cagnes, France.

[5] In the late fourth century B.C. Plato imparted his teaching in an olive grove sacred to Academos, a hero revered by the Greeks. This grove was thus known as the *Academia*, which gave us *academy*. The Academia, where, incidentally, Aristotle studied under Plato and later taught himself, flourished as an institution of learning for some nine centuries, until Justinian, an enemy of all vestiges of paganism, closed it down.

"Olive Grove," acrylic, by Cristina Cini, painted 1996.

Medicine

In earlier times olive oil was used therapeutically in many ways that with our much more extensive pharmacopoeia we do not find interesting today, although it is still favored in some applications. Pliny describes 15 varieties of olives. Roman medicine favored the oil of the bitter wild olive.

Then and now, for instance, warm oil may be introduced to the ear as a remedy for inflammation or to loosen excess wax. Then and now, it is applied as a balm to burns and stings—as formerly to wounds.

We have all become alert to the danger of cholesterol, which can increase the risk of heart disease. Animal fats raise LDL (lipoproteins), called "bad" cholesterol. Polyunsaturated oils lower LDL, but, alas, also turn out to lower high-density lipoproteins (HDL), which is also known as "good" cholesterol, and has the effect of offsetting the "bad" cholesterol.

Fortunately, olive oil and other "good" or mono-unsaturated oils both lower "bad" LDL and simultaneously raise "good" HDL. Olive oil also has another advantage. It contains antioxidants, which offset "bad" free radicals. While it is at it, olive oil contains phyto-estrogens, which counteract bone loss.

Strigil made of bronze, Greece. IV *century* B.C. *They were also made in silver. Museo dell'Olio Fondazione Lungarotti, Torgiano.*

Painted alabaster vase used for perfumed oil. Pasiades, Greece, c.500 B.C.

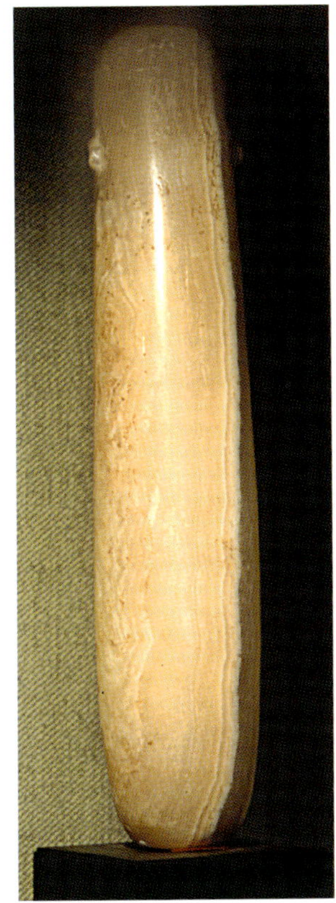

Ointment jar in alabaster, Egypt, 1500–1000 B.C. Museo dell'Olio Fondazione Lungarotti, Torgiano.

Cosmetics

Thousands of years before Christ, olive oil, often perfumed with myrtle, cedar or juniper, was applied to the body. (Oil is a fine base for perfumes.) One thinks of Queen Nausicaa and her companions in the Odyssey anointing themselves after bathing, and indeed, of Odysseus being recognized by his old nurse Eurydea after she bathes and anoints him. In the Iliad the fallen Patroclus is anointed before his funeral rites:

> Achilles ordered his comrades to make preparations for laying out Patroclus' corpse. They kindled fires beneath a large, three-legged cauldron, full of sweet water. Flames wrapped themselves around the belly of the cauldron, and it presently came to a boil. This hot water served to wash the gory corpse, which the layers-out then rubbed with olive oil. After pouring fresh unguents into the wound, they placed Patroclus on a bier and threw over him a thin linen sheet and a white cloak.
>
> Trans. Robert Graves

In ancient Crete olive oil was boiled with rose, cypress or sage to create aromatic oils and unguents, as evidenced by the perfume flasks found in women's graves. Aromatic oil was much favored as a religious offering. Phoenician merchants helped spread the use of scented unguents and there is evidence that they were used by the Etruscans as well.

The Jewish custom of applying perfumed ointment occurs in the story of the woman who pours costly oil on Christ's head on Calvary, and of course in the anointing of His body after the crucifixion.

Patrons of a Roman bath carried a glass, clay or alabaster ampoule containing oil on a delicate chain. Athletes might do likewise. Considered "oriental luxuries," Republican Rome condemned the use of unguents, which then flourished in Imperial Rome.

Interest in perfumed oils persisted through the middle ages and strengthened in the Renaissance. When Catherine de Medici came to France as queen, she was accompanied by her perfumer, a certain Renato.

Incised majolica oil jar for pharmaceutical use, Montepulo Fiorentino, Tuscany, XVII century. Museo dell'Olio, Torgiano.

A wide variety of products made with olive oil: massaging oil, shampoo, shower gel, moisturizing cream. The soaps shown are scented with orange, bran and honey, mint and Mediterranean herbs.

At one time the oldest woman in the world was believed to be Jeanne Calment, of Arles, France. She ascribed her longevity in part to olive oil, which she said she consumed with almost every meal, and rubbed into her skin. "I have only one wrinkle," she declared stoutly: "I'm sitting on it."

This tale reminds us of olive oil's merit as a counter to dry skin, known as far back as the ancient Egyptians and still favored today, in soap, among other applications.

A picturesque old-fashioned middle eastern procedure for making soap is described by a traveler who visited a soap works in an ancient underground cavern with dimly illuminated vaulted stone ceilings. The first step is to boil the waste of the olive-press in caustic soda. With buckets, workers scoop up the thick slurry that forms on the surface of the mixture. A donkey engine hauls up these buckets on a track to the next higher floor. There, on a bed of stone, lie wooden frames, into which the slurry is poured. A few days later it has hardened. A worker wielding a razor-like blade on the end of a stick slices the product lengthwise and crosswise into green cubes. Simple! The blocks are then wrapped in whatever fashion the buyer specifies, and shipped off. You can also make white soap out of actual low-grade olive oil, but it's much more expensive.

The black soap that is ordinarily used in a Turkish *hammam* is composed of 80% olive oil. For modern refined soaps and cosmetics, scents of plants and herbs are added to the oil.

Olive oil soap made with eucalyptus.

The Tree Itself, Producing the Oil

The botanical name for most olives is *Oleacea*. There are 700 varieties. Surprisingly, the lilac, jasmine, and forsythia are kin. The *Olea Europeae* is the commercial species.

Since the olive has been grown for thousands of years, its cultivation has received a vast amount of study. Among the treatises on this subject surviving from ancient Greece are Hesiod's *Works and Days* and Theophrasters' *Historia Plantareum*. In Latin we have Pliny the Elder and the Carthaginian general Mago, whose vast tome was translated by order of the Senate. Cato's *De Agri Cultura* in the second century B.C. urges that the trees be exposed to the west, and specifies the spacing and ditching. Varro's *Rerum Rustiorum* in the first century B.C. offers counsel on arranging the rows of trees, on planting times, suitable fertilizers and insecticides, and harvesting techniques. (He deprecates beating the tree with a stick instead of picking carefully by hand.)

Columella in the first century B.C. urges dry, barren, permeable soil, annual deadsticking, biennial pruning, and triennial fertilizing with manure and oil sediment. In *De Re Rustica* he hailed the olive as "the first among plants." Pliny the Elder's *Historica Naturalis* deplores the huge *latifundia* in favor of intense cultivation in family farms, and recommends picking the fruit before full maturity. Toward the fifth century A.D., Palladius wrote of the olive in his *Opus Agricoltorae*, which was still widely consulted a thousand years later!

From top to bottom, flower, unripe and ripe olives.

The olive's migrations from Egypt to Greece and Italy, and thence to France, Spain and elsewhere, resulted in spontaneous hybridization and conscious selection of the more useful varieties, based on regional conditions and tastes.

Like many other trees, an olive throws out new shoots—"suckers"—from its collar. These can develop foliage and sink new roots, resulting in a process of regeneration. In this way the original tree can carry on for a thousand years. Some famous individual trees are supposed

Olive tree in Sardinia, Italy.

to date back to Roman times! When cut down, the olive throws up suckers from its stump, which bear fruit all over again. Titus Vespasian (A.D. 9–79) is said to have cut down the olive trees in the Garden of Gethsemane, in 70 A.D. The few trees still growing there may be well have regenerated from those Jesus knew. As Pliny wrote, "the olive never dies."

Pruning is essential to sound cultivation. If allowed to grow to great height, the tree becomes difficult to pick.

In general, hard pruning often stimulates the growth of the plant material that remains.

Trees that are not pruned and carefully fertilized are likely to have profuse crops only irregularly. An easy way to propagate the olive is by cutting off branches several feet long and planting them directly in fertilized ground. Sometimes, one can lay branches flat and cover them with earth, after which they may throw up new shoots. Yet another variation is to cut out new buds from the stem and plant them like seeds. One can of course grow trees from seeds in the usual way. As with many tree crops, it is not unusual to graft cultivated seedlings onto separate rootstock.

Olive tree, Apulia, Italy.

The olive tree is only happy with a warm summer and a dry, cold winter, when temperature should fall close to freezing in order to eliminate a troublesome pest, the olive fly. The olive is more productive growing in the plain than on the hillside, and like grapes, thrives in dry, barren soil. It can send down tap roots as deep as 20 feet, searching for water. Nevertheless, it is highly responsive to irrigation and fertilization.

Bark of olive tree, Cagnes, France.

Olive grove, Viterbo, Italy.

Olive tree in Tunisia.

Olive grove in Cagnes, France.

Olive tree in Croatia.

A girl carrying an olive branch, painted on an ancient oil jar. Museo Etrusco Carlo Farina, Orvieto, Italy.

Fragment of ancient mosaic depicting a man picking olives, Tunisia.

Fifty-four workers on a single tree picking olives. These are eating olives, which must be picked by hand. Photographed in 1896. Museo dell'Olio Fondazione Lungarotti, Torgiano.

While ripening, the fruit turns from green and hard to dark and tender. Since the exact time of picking (usually between November and February) determines the stage of ripening that the fruit has reached, the harvester controls the product. The highest quality is usually considered to be the green stage, as with cut flowers. The advantage turns primarily on the unripe fruit's higher concentration of an antioxidant, phenol. This substance offers a natural counter to the olive's aging process, including sunlight and heat. It imparts a peppery flavor and a characteristic aroma that olive-fanciers enjoy. To understand this point of view better, think of a dry martini with a young olive in it: very nice! Now imagine in the drink a mature, soft, fat, almost black fully ripe olive: unsatisfactory. Thus, oil constitutes about a quarter of a ripe fruit.

A comb is used to detach the olives from their branches in the past as today.

About ten pounds of fruit makes one quart of oil, the yield declining as the quality rises. The very highest quality calls for as much as 25 pounds of fruit per quart of oil. This is the so-called *mère goutte*, which is produced just when the olives are crushed, without further pressing.

For the very best results, one picks olives by hand, using a small rake. As when harvesting citrus, several people climb around the tree with ladders. Nets are placed under the tree so that the olives will not hit the ground and split the tender skin. (Such a wound admits bacteria and air, triggering fermentation.) From these nets the olives are conveyed to the press, loosely packed in baskets or bags. After washing, the olives are ground into a paste. In ancient times this was done with stones and mortars, or millstones turned by hand or by donkeys. A museum in Haifa exhibits a stone mortar several thousand years old!

Traditionally, the crushed paste was either put in a earthenware jar or in a screw press to be squeezed. Hot water was poured over the paste while it was kneaded by hand. Being lighter than water, the oil floated to the surface to be skimmed off. The oil then ran down into jars. The more modern procedure is to use an electric or hydraulic crusher, which speeds up the process. (See page 67.)

Olive harvest, Baratti, Italy.

Olives in hemp sacks awaiting transport from the grove to the mill.

A harvesting machine shaking the trees. The olives are collected in the net. La Traversagna, Tuscany.

An olive tree planted in a favorable situation takes some seven years to bear significant fruit, and continues to increase its production—up to 100 kilos a year—as it approaches full growth.[6] The wild olive is a straggly evergreen bush, but the cultivated tree can rise as high as thirty feet or more, with very shallow roots that project out about as far as the width of the tree's aboveground crown. It has thorny branches, and bears small white flowers that are pollinated by the wind. Its evergreen leaves with silvery undersides shimmer in light airs. The tree likes sea breezes. The wood, being very hard and close-grained, is valued by cabinetmakers.

[6] Under Solon the Athenians evolved from aristocracy—rule by the well-born—to rule by the wealthy and powerful. The top class, eligible for the highest offices, were the *Pentakosiomedimnoi*, whose properties produced at least five hundred bushel measures of olive oil, grain, or wine. (Smaller producers enjoyed more limited rights, as did the equestrian and cattle-herding classes, while the old aristocrats also retained some privileges, since Solon liked compromises that worked.)

Transport of the olive, by Giovanni de Grassi and school, 1389–1400. Museo dell'Olio Fondazione Lungarotti, Torgiano.

Spreading and cleaning the olives, photographed in 1896. Museo dell'Olio Fondazione Lungarotti, Torgiano.

Defoliating machines.
La Travesagna, Tuscany.

Detail of an old olive jar. Museo dell'Olio Fondazione Lungarotti, Torgiano.

Late 19th century earthenware jars for olive oil, Calabria, Italy.

Pressing of the olives by Giovanni de Grassi and school, 1389–1400. Museo dell'Olio Fondazione Lungarotti, Torgiano.

Fresco depicting olives being pressed. San Bernardino's miracle, Conversano, 1611. Museo dell'Olio Fondazione Lungarotti, Torgiano.

At Volubilis, Morocco.
(Photos by Lisa Train Pinnington)

In the old town cellar, Orvieto, Italy

Ancient olive presses are found in many Mediterranean archeological sites: this one stands at the seaside at Sabrata, Libya.

An old and still used stone press in Campiglia, Tuscany.

Olive press.
Museo dell'Olio Fondazione Lungarotti, Torgiano.

Wicker pressing plates drip oil while crushing the olives.

An olive mill in Campiglia, Tuscany, photographed in 1949.

A 19th century olive mill in Cagnes, France.

Modern machines used for the production of oil at La Traversagna, Tuscany.

A foal nibbling an olive branch.

Oil in its final stage is being tasted with a piece of celery.

A Visit to La Traversagna

To see the latest in olive growing I decided to visit La Traversagna, the largest single olive grove in Italy. Not far from Lucca, in Tuscany, it extends about one kilometer in each direction, with about 45,000 trees. It is part of the Fontana family's SALOV group, whose main business is blending and selling olive oil. The group includes a dozen or so significant brands, of which the best known is Filippo Berio, in business since the mid 19th century, and today selling 35 million liters per year in fifty countries. The oil produced at La Traversagna is marketed in Italy under the label Sagra, which dates from 1959.

I arrived by invitation on a very hot afternoon in June 2003, having driven by car from Florence. The farmhouse is painted a weathered red, and the working buildings the usual Tuscan yellow. Large terra cotta urns containing flowers stand here and there. The two-story building where visitors are received has a paved courtyard with a couple of metal tables. Inside, the huge, austere reception room is perhaps 100 feet long by 50 feet wide, with an oversized fireplace and a long dining table. On the wall we see faded photographs of earlier times: a costume ball, a visit by dignitaries, some elegantly turned out men standing around a snappy roadster.

After having a snack with a glass or two of white wine, and examining a number of types and brands of the group's oil, we set forth on foot to inspect the trees and the buildings containing the processing machinery. The trees are placed in perfect alignment: vertical, horizontal, and diagonal. Wide paths between the rows permit the passage of the wheeled vehicles used for spraying each tree, collecting the fruit, and cleaning up the area. The trees are relatively young, and pointed, like cypresses, rather than spreading out like oaks. They are topped from time to time so that the higher branches will not become so tall as to be out of convenient reach.

La Traversagna uses sprinkle irrigation, water running through rubber pipes that are strung above ground level from tree to tree. Sprinkle (or drip) irrigation at ground level is known elsewhere, but keeping the pipe off the ground facilitates the cultivation of the area immediately around the tree.

All crops suffer from pests. In Italy, one of the worst is tiny flies that can actually kill most of the fruit. Another one is very small spiders, which can cost a third of the crop.

The plantation grows three different varieties of olives, which, however, are so similar in appearance that they can only be distinguished by a professional. La Traversagna was conceived as a model operation, applying the latest techniques of mechanized production. One is a mechanical harvester that seizes the trunk in huge padded grippers and shakes the whole tree (see page 55), while an upside-down umbrella deployed around the trunk catches the falling fruit. For this to be possible, the tree should grow vertically from a single trunk, rather than spreading out broadly in the traditional fashion. Old-style olive trees grow up in three—or possibly two or four—major branches that are not part of the main trunk, which thus becomes very hard to agitate in this way.

One harvesting machine will collect the olives from three hundred trees per day. At the same time the very top is tapped by men with sticks, to shake off the high crop. The harvested olives are put in a defoliating machine (see page 57), which with a blast of air removes their leaves. Subsequently, there is a second leaf removal process. The olives are then washed and pass to a crusher, including the kernels inside, which are important for nutritional purposes. The resulting slurry goes on to be decanted into oil and water; the residue is separated and can be recycled to fertilize the trees. The oil is stored in stainless steel tanks. To prevent oxidation, they are capped by nitrogen, which is drawn off as the oil level rises.

In a modern mechanized crushing plant, two people can run an operation that in ancient times, using huge stone millstones, might have taken dozens of people.

Since my visit was out of season, there was little movement: neither picking nor processing. At such times the prudent farmer repairs his equipment and gets ready for the next campaign.

The Oil

Some bottles, particularly in France, carry the legend, "*première pression à froid*"—"first cold pressing." This expression dates from the times when you performed an initial pressing, which extracted some of the oil, after which you added hot water—like water-flooding an oil well—and then squeezed out the remainder. Today's powerful hydraulic presses perform the whole operation in a single huge crush, so the French expression becomes meaningless.

New oil, within two or three months of harvesting, is called *olio novello.* Under new European Union standards, eighteen months from harvesting will be the outside period of sale. The oil remains at its prime for up to a year or so after harvesting. Thereafter it declines, becoming dull and finally sour. Olive oil enthusiasts insist that the youthful oil is best. Oxidation starts both to degrade its quality and to blur the distinctiveness of its different sources, in as little as a few months. A superior producer is thus likely to label each container with the date it was harvested. New oil is cloudy, should be shielded from light, and stored at about 14 degrees centigrade, not refrigerated. Oil varies in color from clear through golden to green. The color does not, however, imply quality. Rather it suggests the type of fruit used and sometimes the oil's age. Cloudy oil implies imperfect filtration, or storing at too low a temperature.

Virgin oil is made from fruit that is harvested before fully ripe, peeled, and then pressed without heat. This results in the highest quality. A second pressing produces lower quality, and finally the remaining pulp may be mixed with water to produce oil used in industry. Even beyond that one can add a solvent to produce very poor oil.

The widespread classification of "extra virgin," which seems self-contradictory, like "a little pregnant," is based on a state inspection. Acidity must not exceed one percent (being reduced to 0.8% in the European Union). An eight-person panel performs a blind testing, providing ratings on the good and bad qualities of the oil. Finally, it must have been subjected to no treatments except filtration. Essentially, though, all superior oil is classified "extra virgin."

Oil from 1% up to 3% is simply "virgin." Higher acidity than that can still be called "pure," "light," or "extra light." In Italy all these have the unflattering title of *lampante* or lighting oil. It suffers from a serious defect, namely that it has an unpleasant fatty quality that seems to coat one's mouth.

Many other products may be mixed with olive oil, such as cottonseed and sesame oil, but in the U.S. it must then be labeled as salad oil, not as olive oil. By and large, every country, and indeed every region, produces different flavors of olive and of olive oil. It follows that the dishes and the recipes must also vary greatly.

While the Greeks taught the Romans how to cultivate the olive, it appears that the Etruscans were the first to discover its value in cooking. Before that, however, the other uses—to oil the body, as a cosmetic, and as a source of light—were available.

There has for long been an olive oil/butter frontier in Europe, with America on the butter side of the line. This division is slowly fading as the merits of olive oil, and the disadvantages of fat, become better appreciated.

A fresco depicting a dish of olives from the Topkapi museum, Turkey. (PHOTO BY GIOVANNI ANSELMI)

Cooking With Olive Oil

The enjoyment of fine olive oil, like enjoying fine wine or fine anything, is greatly enhanced by careful, thoughtful tasting. Instead of wolfing it down, the gourmet consumes it slowly and thoughtfully, contemplating its various aspects, and noting the response of various parts of mouth and tongue.

The ancient Romans loved olives, and became extremely knowledgeable about the different varieties and their characteristics. At banquets, for example, guests, blindfolded, might be served olives and olive oil; the knowledgeable guest was expected to identify the different varieties and their origins, just like today's oenophiles.

The usual procedure follows closely that of the wine taster. First note the bouquet, what wine fanciers call the "nose." You pour a little oil into a small glass. Let your hand warm the glass to release the scent. Inhale slowly several times and notice what you detect. Then, the color: As the oil is poured, is it light or dark? Greenish or yellowish?

Finally the taste: Again like the wine taster, but taking the oil in a spoon, roll the oil around the closed mouth. Observe the different tastes as the oil touches different places: perhaps a touch of sweetness at the top of the tongue; acid against the cheek, a touch of bitterness in the "back taste," as wine-drinkers call it. Note the texture that is left behind in the throat: the flavors depend on the type of olive, blends, and vintages, and, of course, from the estate the fruit comes from. The tastes range from peppery to nutty, grassy, fruity, or flowery.

A pleasant Tuscan way to taste olive oil is to place sea salt and grains of black pepper in a shallow bowl, and add the oil. You then dip white bread into the mixture, taste, and wait for the flavors to reveal themselves. A peppery "catch" or aftertaste suggests a newly harvested oil, which gourmets often prefer. Some find that oil on thin slices of potato makes a good basis for serious tasting. Whichever technique you favor (including others that you may discover), when tasting several oils, cleanse the mouth with water, wine, or a slice of apple between each experiment.

An assortment of olives.
(PHOTO BY NICOLAS CHOA)

In the cookbook by Apicius, nearly all the recipes intended for the wealthy called for olive oil. The commonality cooked with lard.

Almost every dish in Mediterranean *haute cuisine* involves a particular olive oil, chosen carefully to harmonize with the dish being served, just as one would do when choosing a wine to accompany a meal. Whenever oil is used, the golden rule is that the oil should enhance the taste of the food, not diminish it. It is generally accepted that for fish, a soft sweet oil, like that from Liguria, is appropriate. For steak and vegetables, one would choose a strong, pungent oil, e.g. from Tuscany. A new perhaps slightly bitter oil is preferred for salads, soups and pasta. Some Italians go one step further, preferring to serve "bruschetta" or "fettuna" bread toward the end of November, precisely when the recently harvested sharp Tuscan oil is available. Hints from an Italian chef will be rewarding: the right oil for *aglio e olio* spaghetti sauce, for tomato, for fish, chicken, and so on. There is little or nothing on olive oil connoisseurship in modern cookbooks, so you'll have to find out for yourself. Give it a try!

In general, the oils from the south of Italy are sweeter and softer than those from the north because they are harvested later in the season.

A very recent invention is aromatic oil. The market supplies oils mixed with all kinds of herbs, hot peppers, juniper seeds, orange, fennel, garlic and even truffles. Olive oil infused with basil makes a delicious condiment, and also with lemon, particularly on fish. Many restaurants now serve a saucer of olive oil to dip delicious warm bread into, instead of butter: an excellent way to appreciate the oil, although the bread will obviously confuse the taste of the oil somewhat. It is amazing that this admirable Italian custom has not been diffused more widely.

Preserving foods in oil—"*sottolio*"—as well as smoking them, has been a traditional method since ancient times, before modern refrigeration. The most common foods preserved in oil today are mushrooms, dried tomatoes, peppers, eggplant, sardines, anchovies and the olives themselves.

Approximately 10% of the olive harvest is used for cured olives. When picked they have a bitter and unpleasant taste. In the late Roman period the unripe olive steeped in brine was prized as a delicacy.

Pickled olives have been found among the buried stores of Pompeii. The olives used for eating must be unblemished and are therefore care-

Pesto is made with oil, parmesan and pecorino cheese, basil, and pinoli from pine acorns.

fully handpicked. There are many ways of curing olives. Every family and every village has its own recipe. A most popular technique to cure ripe, fat black olives is to open a little cut on the side of the fruit, before plunging it in salted water with vinegar and oil to ferment. Small dry olives, having been soaked in water for several days, are dried in the sun. Small brown olives are put in salted water for 24 hours and then in plain water for a month. One can also find large green or black olives stuffed with peppers or anchovies.

What about the olive in a gin cocktail? A vexed subject! Some credit its first composition to Julio Richelieu, a California barkeep, in 1870. The town was called Martinez, so he named his innovation the Martinez cocktail. Others claim that Martini di Arona di Taggia, a bartender in the old Knickerbocker Hotel in New York, stuck an olive in a cocktail that he made for John D. Rockefeller in 1910. Under that theory, his first name was the source of what became the name of the cocktail. Who knows? Both stories could be true, or neither, or others.

The ingredients for "Pappa al Pomodoro" are oil, bread, tomatoes, garlic, basil, capers and olives.

Hors d'ouvre of olives displayed on an olive wood tray containing: large black Greek olives, black cured Italian olives, Kalamate with pimento, small black Spanish olives, white green Greek olives, green large Italian olive. In the center of the tray we have eggplants in oil, herring with pesto, capers and sundried tomatoes. Standing on the marble table, an old ceramic oil container from Calabria, Italy.

Minestrone

Kitchen tools in olive wood.

Red peppers.

Cooked white beans with sage and oil.

Veal dressed with oil, herbs and garlic ready for roasting.

Frittata.

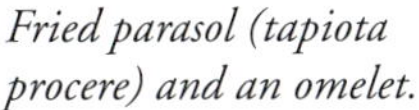

Fried parasol (tapiota procere) and an omelet.

Mushrooms and polenta is cooked with oil, chanterelle (cantarellus cibarius), parsley, onions, polenta and parmesan cheese.

"Bruschetta" is prepared by rubbing garlic on toasted bread that is then covered with tomatoes and basil and smothered with oil.

Tagliatelle with black olive sauce.

Tile table top. Deruta, Umbria, Italy.

The Olive Oil Trade

Around the time of Christ, the olive oil trade was more extensive than even that in wine. Later, Italian production was adversely affected by the disorders of the peninsula and the impact of heavy imports. After the sixth century A.D., wars and invasions led to the decline of Italian cultivation.

The largest market for olive oil is the United States, and the biggest single producer is Spain, with Italy a fairly distant second. French production is minor. Spain is responsible for something over half a million tons of oil, a bit more than a third of total world production. Italy is about a fifth of world production, and Greece somewhat less. However, Italian consumption exceeds production, so to fill the deficit, every year Spain ships something like 200,000 tons to Italy. American consumers, often Italo-Americans, lean toward Italian oil, but it is virtually certain that a lot of Spanish oil finds its way into the U.S. under Italian labels.

In the latter nineteenth century there arrived wave after wave of poor South Italian immigrants into the United States. After the Civil War, black slaves, freed, often left their jobs, being sometimes replaced by Sicilian farm workers. With them they brought their religion and their food—both at home and in the Italian restaurants that proliferated. Also their folkways, including the Sicilian Mafia and the Neapolitan Camorra. The Mafia had originally justified itself as a form of resistance to maintain Sicilian independence, but as usual the criminal activity crowded out the political purpose. In the large American cities with Italian populations the Mafiosi soon got a grip on the protection racket, in which mobsters, having bribed the police, demand tribute from shopkeepers as "protection" from arson and other damage.[7] Also, however, they established oligopolies in the importation of such staples as cheese, artichokes, and olive oil; when Prohibition came in, so did the most prof-

[7] My father, a New York City prosecutor in the early 1900s, traveled to Italy at his own expense to investigate these criminal associations. In his writings he stated that if the United States ever relaxed its then requirement that prospective immigrants bring with them certificates of good conduct from their home parishes, we would be inundated by organized crime. We did, and we are.

itable and violent of those traffics, bootleg liquor. (All economic sanctions, including narcotics, bring criminal control of the commodity in question, accompanied by massive bribery of politicians.)

With Repeal in 1933, the booze business could again flow in legal channels. Some of the bootleggers went in that direction. Some, possessed of an excess of trucks and warehouses, turned, among other trades, to olive oil. The Profacci family went into that activity as its first more or less legitimate business. Of course, there were the usual rough and ready ways of discouraging competition, many of which involved murder and arson. (We see such means today in narcotics, construction, gambling, linen supply, coin vending machines, and garbage disposal.) A long step from the devoted Franciscan fathers in San Diego of earlier centuries!

In our own time, a fraudster named Tino de Angelis perpetrated one of the most complicated crimes ever seen on Wall Street. He filled huge tanks, maintained by an American Express subsidiary, with water topped by a thin coating of olive oil. He then offered these tanks as security for major loans from a number of institutions. When he went bankrupt, there ensued years of litigation, since at the time American

Market in a Medina, Tunisia.

Express was a joint stock company, not a corporation, so the creditors sued not only the tank storage company and its parent, the American Express company, but the shareholders, the brokers who handled the transaction, the insurance companies involved, and each other. This cat's cradle of litigation became so intricate that plaintiffs could scarcely find a law firm not already representing an adverse party. My own brother-in-law, Dean K. Worcester, American Express's lead counsel, had intended to retire, but had to put off that event for years, and indeed to engage a larger law firm to back up his own.

Italian influence in the U.S. olive oil trade is of course extensive, Italians being the principal consumers. Within the Italian community, the mafia is not absent. (Movie enthusiasts may remember that in *The Godfather*, Vito Corleone is blasted by a rival family outside his family olive oil business in Manhattan.) In the 1940s, the Profacci family were perhaps the dominant factor: they held the New York and New Jersey docks in thrall, while other families dealt with other areas. The Bonnano family, notably "Joe Bananas," related by marriage to the Profaccis, was active in the midwest. The present-day Profaccis now engage in normal business, and are proud of their products and their reputation. They deal in Colavita oil, whose headquarters are in Linden, New Jersey, including a 60,000-square-foot warehouse. Colavita assembles oil from various producers, and bottles it in Campobasso, Italy. The family has endowed the Colavita Center for Italian Cooking.

An interesting difference arose between the olive distribution systems in Italy and in the U.S. Sears Roebuck and other catalogue merchandisers spread the consumption of olives where conventional means of distribution might not have penetrated. Many tens of millions of catalogues a year were mailed out at one time. (Today, of course, distribution is effected by the internet and e-mail.) A lot of Sears' success arose because of the confidence consumers had in the reliability of the catalogue merchandisers. In Italy, on the contrary, consumers were more likely to live near a retail distributor, and the postal service was less reliable. Thus a mail-order system of olive sales has never become important in that country.

Spain, not Italy, is both the world's largest producer and largest exporter of olive oil. However, with much higher volume, the dollar value of Spain's exports is about the same as Italy's. Most of Spain's sales are in bulk, while the big money is in the handsomely bottled premium brands, such as Bertolli, that are popular in America.

Italy also imports more oil than it exports, so one cannot help wondering how much of Italy's exported bottled oil is in fact from other sources, not only Spain, but also France, Greece and Portugal. The answer is not established.

Spain's obvious strategy in response to this situation is to develop premium bottled brands of its own, and this it is setting out to do. Studying Italy's marketing achievement, Hojiblanca, Spain's main olive-oil cooperative, realized that while the intrinsic cost of bottled oil is only about 10% over the bulk product, the market value is much higher if the bottles are skillfully designed and presented. Thus, Hojiblanca has signed film star Antonio Banderas to be the "face" of its bottled oil in the export markets, notably the U.S.. Banderas will receive a share of the profits. This effort comes at an opportune moment, since U.S. imports of olive oil are booming, thanks to its vogue in healthier cooking.

Antique oil jar, recovered from the sea.

Maps showing where olive trees were grown in classical times. Museo dell'Olio Fondazione Lungarotti, Torgiano.

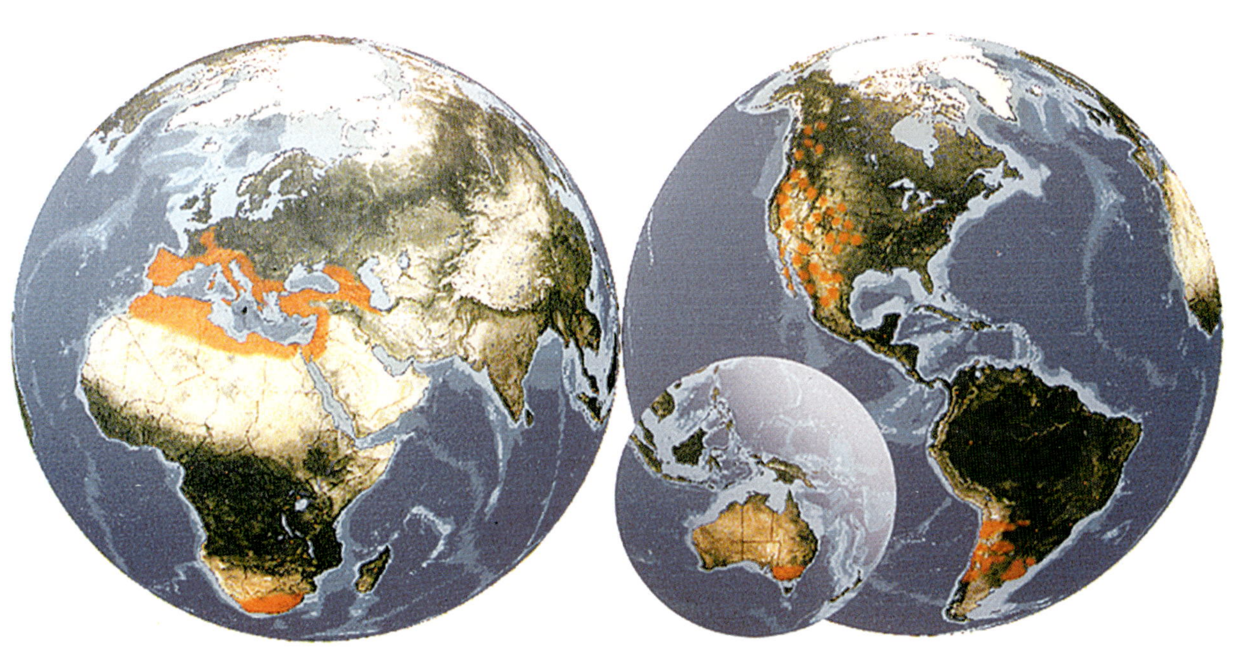

Globes showing (in orange color) where olive trees grow today. Museo dell'Olio Fondazione Lungarotti, Torgiano.

Conclusion

An Arabic riddle: "Our servant is green, her children are white, then grow black. Who is she?"

The reader will not be mystified, but perhaps somebody else will.

Detail of tiles in the garden of Fronteira, Lisbon, Portugal.

Appendix
Superior Olives

Italy

1. **Cerignola**. Very large and green; grown around Bari.
2. **Gaeta**. Purple-brownish, from central Italy. Excellent on pasta or pizza.
3. **Ligurian**. Black, high in oil; sweet, delicate taste. Both for oil and to eat. Grown in the hills around Genoa.

France

1. **Nicoise**. Essential in salade nicoise; black, very delicate flavor.
2. **Nyons**. The favorite for regular consumption of some connoisseurs. Black, with a rich, full flavor. Be sure of your supplier, so that you don't get North African substitutes.
3. **Picholine**. Green, with a nutty flavor, from the south of France.

Spain

1. **Gordal**. Sometimes called "queen": large, green, meaty.
2. **Manzanilla**. Brownish-green; smaller and nuttier than gordal.

Greece

1. **Amfissa**. Purple-black, from the Delphi region. Soft and sweet.
2. **Kalamatas**. The standard Greek olive, from the western end of the Peloponnesus. Purple-black with a pointed shape. Connoisseurs greatly prefer those that are hand-picked.

3. **Nafplion**. From the eastern Peloponessus.
4. **Thassos**. Meaty, with an intense flavor.

Middle East and Israel

1. **Souri**.

North Africa

1. **Sahli**. Small, black, easy to eat.

USA

1. **Mission**. Slightly bitter. The original American olive, brought up by the Franciscans from Mexico.
2. **Sevillano**. Large, green, meaty.

New Zealand

1. **Moutere Grove**. Soft, fruity, and peppery; expensive.

South America

1. **Los Ranchos**, Uruguay. Grassy and soft.

John Train is author of twenty books and several hundred columns. Chairman of an investment firm, he has received part-time appointments from three President.

Mark E. Smith is specialized in travel reportage, fine art, nude, archeological and architectural photography.